Brain Health Diet Simple Guide for Beginners

Understanding How Your Diet Affects Brain Function

By

Lomond Kenzie

Table of Contents

CHAPTER 1

Introduction

1.1 Understanding the Importance of Brain Health

The human brain, often referred to as the body's command center, is an intricately complex and indispensable organ. It governs our thoughts, emotions, memories, and physical movements, allowing us to interact with the world around us and to experience life to its fullest. Understanding the importance of brain health is not merely an academic pursuit; it is a matter of personal well-being and quality of life.

Your brain health affects every aspect of your daily existence. It determines your ability to learn, to solve problems, to make decisions, and to adapt to new challenges. It also profoundly influences your emotional well-being, impacting your mood, stress levels, and even your relationships. Moreover, maintaining good brain health is essential as we age, as it can help stave off cognitive decline and neurodegenerative diseases such as Alzheimer's and Parkinson's.

Imagine your brain as a finely tuned machine. Just as you would ensure that a high-performance car receives the best fuel and maintenance to function optimally, your brain requires proper care and nourishment to operate at its peak. Brain health is about more than just the absence of disease; it is about maximizing your cognitive abilities and preserving your mental vitality throughout your life.

1.2 How Your Diet Affects Brain Function

Now that we've established the importance of brain health, let's explore a key factor that significantly influences it—your diet. It's not an exaggeration to say that "you are what you eat" when it comes to your brain. Your dietary choices have a profound impact on how your brain functions, and this is a relationship that should not be underestimated.

Consider your brain as a high-performance engine that requires a specific type of fuel to function optimally. Just as a sports car needs premium gasoline to achieve top speed and efficiency, your brain relies on a balanced and nutritious diet to perform at its best.

The foods you consume are rich sources of nutrients, vitamins, minerals, and other bioactive compounds that directly affect brain function. For example, omega-3 fatty acids, commonly found in fatty fish like salmon, are known to support brain health by enhancing cognitive functions and reducing the risk of cognitive decline. Antioxidants, abundant in fruits and vegetables, protect your brain from oxidative stress and inflammation, which can contribute to neurodegenerative diseases.

Moreover, your diet can influence neurotransmitter production, affecting your mood, memory, and overall cognitive abilities. Certain foods, like those containing tryptophan, can help boost serotonin levels, contributing to improved mood and emotional well-being.

Conversely, a poor diet, high in processed foods, sugar, saturated fats, and artificial additives, can have detrimental effects on brain function. It can lead to cognitive impairments, mood swings, and increased susceptibility to mental health issues.

The relationship between your diet and brain function is intricate and profound. This guide will delve deeper into this connection, providing you with insights, strategies, and practical tips to optimize your diet for the benefit of your brain health. As we embark on this journey, remember that the choices you make at the

dinner table have the potential to
shape your cognitive destiny.

CHAPTER 2

The Basics of Brain Health

2.1 What Is Brain Health?

Brain health is a comprehensive concept that encompasses the overall well-being and functionality of your brain. It is not merely the absence of disease or cognitive decline; rather, it's about optimizing your brain's performance and resilience at every stage of life.

Think of brain health as a state in which your brain operates at its best, allowing you to:

- **Think clearly:** Your cognitive abilities, including memory, problem-solving, and decision-making, are sharp and efficient.

- **Feel emotionally balanced:** You experience stable moods and are better equipped to manage stress and emotional challenges.

- **Maintain physical coordination:** Your brain effectively communicates with your body to ensure smooth movements and physical well-being.

- **Adapt to change:** You have the cognitive flexibility to learn, adapt, and grow throughout your life.

- **Prevent cognitive decline:** You take proactive steps to

reduce the risk of
neurodegenerative diseases and
maintain mental vitality as you
age.

Achieving and sustaining brain health involves a holistic approach that includes not only your diet but also lifestyle factors like physical activity, sleep, mental stimulation, and social connections. It's a lifelong commitment to nurturing your brain's potential and safeguarding it against potential threats.

2.2 Common Brain Health Challenges

While the concept of brain health is inspiring, many people encounter common challenges that can hinder their journey towards optimal cognitive well-being. Understanding these challenges is the first step in overcoming them. Here are some of the most prevalent ones:

- **Stress:** Modern life is often accompanied by high levels of stress, which can have a detrimental impact on brain health. Chronic stress can lead to inflammation, impair memory, and increase the risk of anxiety and depression.

- **Sedentary Lifestyle:** Lack of physical activity not only affects your physical health but also your brain. A sedentary lifestyle can contribute to cognitive decline, as regular exercise is known to enhance brain function and protect against neurodegenerative diseases.

- **Poor Diet:** As discussed in the previous section, diet plays a crucial role in brain health. Consuming an unhealthy diet, rich in processed foods and lacking in essential nutrients, can impair cognitive function and increase the risk of brain-related disorders.

- **Lack of Mental Stimulation:** Mental stagnation can be a significant challenge. Failing to engage in intellectually stimulating activities, such as reading, puzzles, or learning new skills, may result in cognitive decline over time.

- **Social Isolation:** Loneliness and social isolation can have adverse effects on brain health. Human connection and interaction are essential for

maintaining emotional and
cognitive well-being.

- **Sleep Deprivation:** Inadequate
  sleep can impair memory,
  concentration, and emotional
  regulation. Chronic sleep
  deprivation is associated with a
  higher risk of
  neurodegenerative diseases.

2.3 The Role of Nutrition in Brain Health

Nutrition plays a pivotal role in brain health, influencing its development, function, and long-term well-being. we will explore how the food you consume directly impacts your brain and discuss key nutrients and dietary patterns that can promote optimal cognitive function.

Nutrients for Brain Health:

1. **Omega-3 Fatty Acids:** These essential fats, found abundantly in fatty fish (like salmon, mackerel, and sardines), flaxseeds, and walnuts, are critical for brain health. Omega-3s, particularly EPA (eicosapentaenoic acid) and DHA (docosahexaenoic acid), support brain cell structure and communication, reduce inflammation, and may lower the risk of cognitive decline.

2. **Antioxidants:** Fruits and vegetables are rich sources of antioxidants like vitamins C and E, as well as phytochemicals such as flavonoids. These compounds help protect your brain from oxidative stress, which can

contribute to age-related cognitive decline and neurodegenerative diseases.

3. **Vitamins and Minerals:** Certain vitamins and minerals are vital for brain health. Vitamin B complex (B6, B9, and B12) supports neurotransmitter production, while minerals like magnesium and zinc are essential for cognitive function. Leafy greens, lean meats, whole grains, and nuts are excellent sources of these nutrients.

4. **Proteins:** Protein-rich foods like lean meats, poultry, fish, beans, and tofu provide amino acids necessary for neurotransmitter synthesis. These neurotransmitters, such as serotonin and dopamine,

influence mood and cognitive
function.

5. **Carbohydrates:** Complex
   carbohydrates like whole
   grains, fruits, and vegetables
   are the brain's primary source
   of energy. They provide a
   steady supply of glucose, the
   brain's preferred fuel, which
   supports concentration and
   mental alertness.

6. **Phospholipids:** Phospholipids,
   such as choline, are essential
   for cell membrane structure and
   function in the brain. Eggs,
   fish, and lean meats are good
   sources of choline.

7. **Hydration:** Proper hydration is
   vital for brain health. Even mild
   dehydration can impair
   cognitive function, affecting

memory, attention, and decision-making. Aim to drink enough water throughout the day.

Dietary Patterns for Brain Health:

1. **Mediterranean Diet:** This diet, rich in fruits, vegetables, whole grains, fish, olive oil, and nuts, is associated with reduced risk of cognitive decline and better cognitive function in older adults. It emphasizes healthy fats, antioxidants, and anti-inflammatory foods.

2. **DASH Diet (Dietary Approaches to Stop Hypertension):** This diet promotes heart health and may also benefit the brain. It encourages the consumption of fruits, vegetables, whole grains,

lean proteins, and low-fat dairy while reducing sodium intake.

3. **Mind Diet:** The MIND diet is a combination of the Mediterranean and DASH diets, with a specific focus on foods that are particularly beneficial for brain health. It includes green leafy vegetables, berries, nuts, and fish.

4. **Intermittent Fasting:** Some research suggests that intermittent fasting may have cognitive benefits by promoting brain health and reducing the risk of neurodegenerative diseases. However, it's essential to consult with a healthcare professional before starting any fasting regimen.

5. **Anti-Inflammatory Diet:**
   Chronic inflammation is linked
   to cognitive decline and
   neurodegenerative diseases. An
   anti-inflammatory diet
   emphasizes foods like turmeric,
   ginger, berries, and fatty fish
   that can reduce inflammation.

These nutrients and dietary patterns
into your eating habits can promote
brain health and support cognitive
function throughout your life.

CHAPTER 3

Building Blocks of a Brain-Boosting Diet

3.1 Essential Nutrients for Brain Health

To support your brain's well-being, it's essential to provide it with the right nutrients. Let's break down these nutrients into three categories and discuss their significance for brain health:

Omega-3 Fatty Acids

Omega-3 fatty acids are a type of polyunsaturated fat that is exceptionally beneficial for brain health. They are a vital structural component of brain cell membranes and play a crucial role in maintaining the flexibility and integrity of these membranes. Two primary types of omega-3 fatty acids are essential for brain health:

- **EPA (Eicosapentaenoic Acid):** EPA is known for its anti-inflammatory properties. It helps reduce inflammation in the brain, which is associated with cognitive decline and neurodegenerative diseases like Alzheimer's.

- **DHA (Docosahexaenoic Acid):** DHA is the most abundant omega-3 fatty acid in the brain and is essential for various cognitive functions. It supports memory, learning, and overall brain development. Adequate DHA intake is particularly crucial during pregnancy and early childhood.

Sources of Omega-3 Fatty Acids:

- Fatty fish like salmon, mackerel, sardines, and trout

- Flaxseeds and flaxseed oil

- Walnuts

- Chia seeds

- Hemp seeds

Incorporating these omega-3-rich foods into your diet can provide essential fatty acids that support brain health.

Antioxidants

Antioxidants are compounds that protect your brain from oxidative stress, a process that can lead to cell damage and inflammation. By reducing oxidative stress, antioxidants help maintain cognitive function and lower the risk of neurodegenerative diseases.

Common antioxidants important for brain health include:

- **Vitamin C:** Found in citrus fruits, strawberries, and bell peppers, vitamin C helps protect brain cells from damage caused by free radicals.

- **Vitamin E:** Nuts, seeds, and vegetable oils are good sources of vitamin E, which acts as a powerful antioxidant in the brain.

- **Flavonoids:** These plant compounds, found in berries, dark chocolate, and tea, have anti-inflammatory and antioxidant properties that benefit brain health.

- **Polyphenols:** Present in foods like green tea, red wine, and olive oil, polyphenols have been associated with improved cognitive function and a reduced risk of cognitive decline.

- **Curcumin:** Found in turmeric, curcumin is known for its anti-inflammatory and antioxidant effects, potentially reducing the risk of brain diseases.

Vitamins and Minerals

Several vitamins and minerals are essential for maintaining cognitive function and overall brain health:

- **B Vitamins (B6, B9, B12):** These vitamins play a role in neurotransmitter synthesis and brain cell maintenance. They can be found in leafy greens, fortified cereals, and lean meats.

- **Magnesium:** Magnesium supports nerve function and is involved in memory and learning. It can be obtained from nuts, seeds, whole grains, and leafy greens.

- **Zinc:** Zinc is important for maintaining cognitive function, and it can be found in lean meats, beans, and nuts.

- **Iron:** Iron is necessary for oxygen transport to the brain. Good sources include lean meats, poultry, fish, and fortified cereals.

Ensuring a balanced intake of these vitamins and minerals through a diverse diet is essential for supporting your brain's daily functions and protecting against cognitive decline.

3.2 The Importance of Hydration

Hydration is a fundamental aspect of overall health, and it plays a crucial role in brain health. we'll explore the significance of staying adequately hydrated for optimal cognitive function and well-being.

Why Hydration Matters for Brain Health:

1. **Brain Function:** The brain is made up of approximately 75% water. Adequate hydration ensures that the brain receives

the necessary fluids to maintain proper function. Even mild dehydration can impair cognitive performance, affecting memory, attention, and decision-making.

2. **Focus and Concentration:** Dehydration can lead to difficulties in focusing and concentrating on tasks. Staying hydrated helps sustain mental alertness and the ability to stay on task.

3. **Mood and Emotional Well-Being:** Dehydration can affect mood stability, potentially leading to irritability and increased stress levels. Maintaining proper hydration can help regulate mood and emotional well-being.

4. **Memory and Learning:**
    Hydration is essential for
    optimal memory and learning.
    Dehydrated individuals may
    experience difficulties in
    recalling information and
    retaining new knowledge.

5. **Energy Levels:** Fatigue is a
    common symptom of
    dehydration. When you're well-
    hydrated, you're more likely to
    feel energetic and motivated,
    which can positively impact
    your overall cognitive
    performance.

Tips for Staying Hydrated:

- **Drink Water Regularly:**
  Make it a habit to sip water
  throughout the day, even if you
  don't feel thirsty. Waiting until

you're thirsty can sometimes indicate mild dehydration.

- **Monitor Fluid Loss:** Pay attention to factors that can increase fluid loss, such as exercise, hot weather, or illness. In these situations, increase your fluid intake accordingly.

- **Incorporate Hydrating Foods:** Some foods, like watermelon, cucumber, and oranges, have high water content and can contribute to your hydration.

- **Limit Dehydrating Beverages:** Reduce the consumption of beverages that can dehydrate you, such as excessive caffeine and alcohol.

- **Use a Hydration App or Bottle:** Consider using a

smartphone app or a marked water bottle to track and manage your daily water intake.

3.3 Balancing Macronutrients

Balancing macronutrients—carbohydrates, proteins, and fats—is essential for a well-rounded, brain-boosting diet. Each macronutrient serves unique functions in supporting brain health and overall well-being:

1. Carbohydrates: Carbohydrates are the brain's primary source of energy. They are broken down into glucose, which fuels brain cells and provides energy for cognitive functions. Complex carbohydrates, such as whole grains, fruits, and vegetables, provide a steady supply of glucose, promoting mental alertness and sustained focus.

2. Proteins: Proteins are crucial for brain function as they provide amino acids needed for neurotransmitter synthesis. Neurotransmitters like serotonin and dopamine play significant roles in mood regulation and cognitive processes. Lean meats, poultry, fish, beans, and tofu are excellent sources of protein.

3. Fats: Fats are essential for brain health, particularly omega-3 fatty acids like EPA and DHA. These fats support brain cell structure, reduce inflammation, and contribute to cognitive function. Healthy sources of fats include fatty fish, nuts, seeds, avocados, and olive oil.

Balancing macronutrients in your meals helps ensure that your brain

receives a steady supply of energy
and the building blocks it needs for
optimal function. Aim for a balanced
diet that includes a variety of foods
from each macronutrient group to
support both your physical and mental
well-being.

CHAPTER 4

Foods for a Healthy Brain

4.1 Brain-Boosting Superfoods

Certain foods, often referred to as "superfoods," are packed with nutrients and antioxidants that offer exceptional benefits for brain health. Here are some brain-boosting superfoods to include in your diet:

1. Blueberries: Blueberries are rich in antioxidants, particularly flavonoids, which can help improve memory and delay cognitive aging.

2. Leafy Greens: Spinach, kale, and other leafy greens are excellent

sources of vitamins (such as folate
and vitamin K) and antioxidants,
which support cognitive function.

3. Avocado: Avocado provides
healthy monounsaturated fats that
support healthy blood flow, and its
folate content may help improve
cognitive function.

4. Turmeric: The active compound in
turmeric, curcumin, has anti-
inflammatory and antioxidant
properties, potentially reducing the
risk of brain diseases.

5. Broccoli: Broccoli is high in
vitamin K and choline, which can
help improve cognitive function and
memory.

6. Dark Chocolate: Dark chocolate
with high cocoa content contains
flavonoids that may boost memory,
mood, and cognitive function.

7. Pumpkin Seeds: These seeds are rich in magnesium, iron, zinc, and antioxidants, which support brain health.

4.2 Incorporating Fish into Your Diet

Fatty fish, such as salmon, mackerel, sardines, and trout, are exceptionally beneficial for brain health due to their high omega-3 fatty acid content, particularly EPA and DHA. These fatty acids have numerous cognitive benefits:

- **Memory:** Omega-3 fatty acids support memory and may help prevent age-related memory decline.

- **Mood:** They can improve mood and reduce the risk of depression and anxiety.

- **Brain Development:** DHA is crucial for brain development in infants and children.

- **Protection:** Omega-3s have anti-inflammatory and antioxidant properties that protect the brain from oxidative stress and inflammation, which are linked to neurodegenerative diseases.

fish into your diet, ideally at least twice a week, can provide a significant brain health boost. If you're not a fan of fish, consider fish oil supplements as an alternative source of omega-3 fatty acids.

4.3 Nuts, Seeds, and Brain Health

Nuts and seeds are nutrient-dense foods that offer various benefits for brain health:

1. Walnuts: Walnuts are rich in omega-3 fatty acids, antioxidants, and vitamin E. They support brain function and may help lower the risk of neurodegenerative diseases.

2. Almonds: Almonds are high in vitamin E, which has been associated with better cognitive function in older adults. They also provide healthy fats and protein.

3. Flaxseeds: Flaxseeds are an excellent source of alpha-linolenic acid (ALA), a type of omega-3 fatty acid. While not as potent as EPA and

DHA, ALA can still benefit brain health.

4. Chia Seeds: Chia seeds are packed with omega-3s, fiber, and antioxidants. They can help maintain cognitive function and provide sustained energy.

5. Pumpkin Seeds: Pumpkin seeds are rich in magnesium, iron, zinc, and antioxidants. These minerals support brain function and overall well-being.

nuts and seeds into your diet as snacks, in smoothies, or as toppings for salads and yogurt to harness their brain-boosting benefits. Be mindful of portion sizes, as nuts and seeds are calorie-dense foods.

4.4 Leafy Greens and Vegetables

Leafy greens and vegetables are nutritional powerhouses that provide a wide array of vitamins, minerals, and antioxidants beneficial for brain health. Here's why they are essential:

- **Vitamins and Minerals:** Leafy greens like spinach, kale, and Swiss chard are rich in brain-boosting nutrients like folate, vitamin K, and magnesium. These nutrients are vital for cognitive function and memory.

- **Antioxidants:** Vegetables, particularly colorful ones like bell peppers, carrots, and tomatoes, are abundant in antioxidants, such as vitamin C and beta-carotene. These antioxidants protect brain cells from oxidative stress and inflammation.

- **Fiber:** Fiber-rich vegetables support overall gut health, which has a strong connection to brain health. A healthy gut microbiome can positively influence cognitive function and mood.

- **Anti-Inflammatory Properties:** Some vegetables, such as broccoli and Brussels sprouts, contain compounds with anti-inflammatory properties that may reduce the risk of neurodegenerative diseases.

To include more leafy greens and vegetables in your diet, consider adding them to salads, stir-fries, omelets, or as side dishes. Experiment with various cooking methods to make them more appealing and flavorful.

4.5 The Role of Berries

Berries, including blueberries, strawberries, raspberries, and blackberries, are often referred to as "brain berries" due to their remarkable brain-boosting properties:

- **Antioxidants:** Berries are packed with antioxidants, particularly flavonoids. These compounds can help improve memory, delay cognitive aging, and protect against brain cell damage.

- **Inflammation Reduction:** The antioxidants in berries have anti-inflammatory effects that may reduce the risk of neurodegenerative diseases.

- **Brain Function Enhancement:** Regular consumption of berries has been associated with improved cognitive function and better performance in memory-related tasks.

Berries can be enjoyed fresh, frozen, or as additions to cereals, yogurt, smoothies, and desserts. Their natural sweetness makes them a delicious and nutritious treat.

4.6 Whole Grains for Cognitive Function

Whole grains, such as brown rice, quinoa, oats, and whole wheat, are rich sources of complex

carbohydrates that provide steady
energy for your brain. Here's how
they support cognitive function:

- **Stable Blood Sugar:** The slow release of glucose from whole grains helps maintain stable blood sugar levels. This is essential for sustaining mental focus and preventing energy crashes.

- **Fiber:** Whole grains are rich in dietary fiber, which supports gut health and influences brain health through the gut-brain connection.

- **B Vitamins:** Whole grains contain B vitamins like B6 and niacin, which are essential for neurotransmitter production and cognitive function.

- **Antioxidants:** Some whole grains, such as quinoa, contain antioxidants that protect brain cells from oxidative stress.

To incorporate more whole grains into your diet, choose whole grain bread, pasta, and rice over refined versions. Experiment with whole grains like quinoa and bulgur in salads and grain bowls. Whole grain oats also make a nutritious and filling breakfast option.

CHAPTER 5

Foods to Limit or Avoid

5.1 The Impact of Sugary Foods

Excessive consumption of sugary foods and beverages can have detrimental effects on brain health and cognitive function:

- **Blood Sugar Spikes:** Consuming sugary foods leads to rapid spikes in blood sugar levels, followed by crashes. These fluctuations can impair concentration, mood stability, and overall cognitive performance.

- **Inflammation:** High sugar intake is associated with chronic inflammation, which can increase the risk of neurodegenerative diseases like Alzheimer's.

- **Cognitive Decline:** Research suggests a link between high sugar intake and an increased risk of cognitive decline and dementia in older adults.

- **Addiction-Like Response:** Sugar can trigger a reward system in the brain, leading to cravings and overconsumption, similar to addictive substances.

To reduce the impact of sugary foods on brain health:

- Limit the consumption of sugary snacks, candies,

pastries, and sugary beverages like soda and fruit juices.

- Choose whole fruits instead of fruit juices to benefit from fiber, which can slow the absorption of sugar.

- Read food labels to identify added sugars in processed foods and opt for products with lower sugar content.

5.2 Processed Foods and Brain Health

Processed foods, including fast food, convenience meals, and heavily packaged snacks, often contain a range of additives, preservatives, unhealthy fats, and excessive salt and sugar. These factors can negatively affect brain health in several ways:

- **Inflammation:** Processed foods are typically high in unhealthy fats, such as trans fats and saturated fats, which can promote inflammation in the brain.

- **Cognitive Impairment:** Diets rich in processed foods have been linked to cognitive impairment and an increased risk of neurodegenerative diseases.

- **Nutrient Deficiency:** Consuming processed foods may lead to nutrient deficiencies as they often lack the essential vitamins, minerals, and antioxidants required for optimal brain function.

- **Gut Health:** Processed foods are generally low in fiber and can negatively impact gut health. An unhealthy gut microbiome can influence brain health through the gut-brain connection.

To promote brain health, consider these strategies:

- Minimize your intake of processed foods, fast food, and sugary snacks.

- Opt for whole, unprocessed foods like fruits, vegetables, lean proteins, whole grains, and nuts.

- Cook at home whenever possible, as homemade meals allow you to control the quality of ingredients and reduce the consumption of processed foods.

- Be mindful of food labels and ingredient lists. Avoid products with a long list of additives, preservatives, and artificial flavorings.

- Choose healthier cooking methods, such as baking, grilling, steaming, or sautéing, instead of frying.

Reducing the consumption of sugary and processed foods, you can protect and enhance your brain health, improving cognitive function and reducing the risk of neurodegenerative diseases in the long term.

5.3 Reducing Saturated and Trans Fats

Saturated and trans fats are unhealthy fats that can have a detrimental impact on both cardiovascular and brain health. Reducing your intake of these fats is crucial for maintaining cognitive function and overall well-being.

Saturated Fats:

Saturated fats are primarily found in animal products and some tropical oils like coconut oil and palm oil. Here's how they can affect brain health:

- **Inflammation:** A diet high in saturated fats can promote inflammation, which is linked to cognitive decline and neurodegenerative diseases.

- **Blood Flow:** Saturated fats can contribute to the buildup of plaque in blood vessels, reducing blood flow to the brain and potentially impairing cognitive function.

- **Cognitive Decline:** Some studies have shown that diets high in saturated fats are associated with a higher risk of cognitive decline and dementia.

To reduce saturated fat intake:

- Choose lean cuts of meat and poultry, and trim visible fat before cooking.

- Use healthier cooking methods like baking, grilling, or steaming instead of frying.

- Limit consumption of high-fat dairy products, such as full-fat cheese and butter. Opt for low-fat or fat-free alternatives.

- Replace saturated fats with healthier fats like monounsaturated and polyunsaturated fats found in olive oil, nuts, seeds, and fatty fish.

Trans Fats:

Trans fats are artificial fats created through a process called hydrogenation, which turns liquid oils into solid fats. They are commonly found in fried foods, baked goods, and many processed and packaged snacks. The impact of trans fats on brain health includes:

- **Inflammation:** Trans fats promote inflammation in the body and can lead to oxidative stress, potentially damaging brain cells.

- **Cardiovascular Health:** Trans fats are known to increase the risk of heart disease, which can have indirect effects on brain health due to reduced blood flow and oxygen supply to the brain.

- **Cognitive Decline:** Some studies suggest that diets high in trans fats may be associated with a higher risk of cognitive impairment and Alzheimer's disease.

To reduce trans-fat intake:

- Check food labels for partially hydrogenated oils, which indicate the presence of trans fats, and avoid products containing them.

- Limit consumption of fried foods, commercial baked goods, and processed snacks, as these often contain trans fats.

- Choose healthier fats for cooking and baking, such as olive oil, canola oil, or avocado oil.

reducing your intake of saturated and trans fats and replacing them with healthier fats, you can protect your brain health, support cognitive function, and lower the risk of cognitive decline and neurodegenerative diseases.

CHAPTER 6

Meal Planning for Brain Health

Meal planning is a crucial aspect of maintaining a brain-boosting diet. we will explore strategies for designing balanced, brain-boosting meals, practicing portion control and mindful eating, and maintaining regular eating patterns to support cognitive function and overall well-being.

6.1 Designing Balanced Brain-Boosting Meals

Balanced meals provide a steady supply of nutrients, energy, and essential components that support brain health. Here's how to design brain-boosting meals:

- **Incorporate a Variety of Foods:** Include a mix of fruits, vegetables, lean proteins, whole grains, and healthy fats in each meal. This diversity ensures you get a wide range of nutrients beneficial for brain health.

- **Prioritize Omega-3s:** Include fatty fish like salmon, mackerel, or sardines at least twice a week to ensure you receive essential omega-3 fatty acids.

- **Lean Protein:** Incorporate lean sources of protein like poultry, tofu, beans, and legumes. Protein provides amino acids needed for neurotransmitter production.

- **Colorful Vegetables:** Aim for a colorful plate with a variety of vegetables, as different colors indicate different antioxidants and nutrients.

- **Whole Grains:** Choose whole grains like brown rice, quinoa, and whole wheat pasta over refined grains to provide steady energy and fiber.

- **Healthy Fats:** Include sources of healthy fats like olive oil, nuts, seeds, and avocados to support brain cell structure and reduce inflammation.

- **Hydration:** Don't forget about hydration. Water is essential for brain function, so drink water regularly throughout the day.

6.2 Portion Control and Mindful Eating

Portion control and mindful eating are essential practices for maintaining a healthy brain-boosting diet:

- **Portion Control:** Pay attention to portion sizes to avoid overeating. Use smaller plates and bowls to help control portions and prevent excessive calorie intake.

- **Mindful Eating:** Practice mindful eating by savoring each bite, chewing slowly, and paying attention to your body's hunger and fullness cues. This helps prevent overindulgence and allows you to enjoy your meals more fully.

- **Avoid Distractions:** Eating while distracted, such as watching TV or working on a computer, can lead to overeating. Try to focus on your meal and savor the flavors and textures.

- **Listen to Your Body:** Eat when you're hungry and stop when you're satisfied. Avoid eating out of boredom or emotional reasons.

6.3 The Importance of Regular Eating Patterns

Maintaining regular eating patterns, including consistent meal times, can benefit brain health in several ways:

- **Stable Blood Sugar:** Eating at regular intervals helps maintain stable blood sugar levels, which are essential for sustained energy and cognitive function.

- **Consistent Energy:** Regular meals provide your brain with a steady supply of nutrients and energy, preventing energy crashes and mood swings.

- **Cognitive Performance:** Consistency in eating patterns can support consistent cognitive performance, helping you stay focused and alert throughout the day.

- **Appetite Regulation:** Regular eating patterns can help regulate appetite and prevent overeating during later meals.

To establish regular eating patterns:

- Aim for three balanced meals a day with healthy snacks in between if needed.

- Try to eat at roughly the same times each day to establish a routine.

- Listen to your body's hunger cues, and don't skip meals.

practicing portion control, mindful eating, and maintaining regular eating patterns, you can optimize your brain-boosting diet and support cognitive function, mood stability, and overall brain health.

CHAPTER 7

Practical Tips for Implementing the Brain Health Diet

To successfully implement a brain health diet, it's essential to make practical changes in your daily life. we'll provide tips for grocery shopping, preparing brain-healthy meals, and dining out while prioritizing brain health.

7.1 Grocery Shopping for Brain Health

1. **Make a Shopping List:** Plan your meals for the week and

create a shopping list. This helps you focus on purchasing brain-boosting foods and reduces the likelihood of buying unhealthy options.

2. **Shop the Perimeter:** In most grocery stores, the perimeter is where you'll find fresh produce, lean proteins, and whole foods. Spend more time shopping in these areas and less time in the processed food aisles.

3. **Read Labels:** When buying packaged foods, read labels carefully. Look for products with minimal added sugars, unhealthy fats, and artificial additives.

4. **Buy in Bulk:** Purchase brain-healthy staples like whole grains, legumes, nuts, and seeds

in bulk to save money and reduce packaging waste.

5. **Choose Seasonal and Local Produce:** Seasonal fruits and vegetables are often fresher and more affordable. Buying local produce supports both your health and the local economy.

6. **Limit Impulse Purchases:** Avoid buying snacks or unhealthy foods on impulse. Stick to your shopping list and stay mindful of your brain health goals.

7. **Opt for Frozen and Canned Foods:** Frozen fruits and vegetables, as well as canned fish like tuna and salmon, are convenient options that can be just as nutritious as fresh when

chosen without added sugars or
salt.

7.2 Preparing Brain-Healthy Meals

1. **Meal Prep:** Set aside time for meal prep each week. Prepare ingredients in advance, such as chopping vegetables, so you can assemble meals quickly during busy days.

2. **Cook at Home:** Homemade meals give you full control over ingredients and cooking methods. Experiment with brain-boosting recipes and try new dishes that incorporate superfoods.

3. **Batch Cooking:** Cook large batches of brain-healthy meals

and freeze portions for future use. This saves time and ensures you have nutritious options readily available.

4. **Limit Cooking Oils:** When cooking, use healthy oils like olive oil or avocado oil in moderation. Avoid excessive use of cooking fats, which can add unnecessary calories.

5. **Use Herbs and Spices:** Enhance the flavor of your dishes with herbs and spices like turmeric, rosemary, and cinnamon. These seasonings provide both taste and brain-boosting benefits.

6. **Incorporate Brain Foods:** Include brain-boosting foods like fatty fish, leafy greens,

berries, and whole grains in your meals regularly.

7.3 Dining Out with Brain Health in Mind

1. **Review the Menu:** Before ordering, carefully review the menu for healthier options. Many restaurants now offer dishes with nutritional information, helping you make informed choices.

2. **Choose Lean Proteins:** Opt for dishes that feature lean protein sources like grilled chicken, fish, or tofu. Avoid fried or heavily battered options.

3. **Customize Your Order:** Don't hesitate to ask for substitutions or modifications to make your

meal healthier. For example,
request a side salad instead of
fries.

4. **Control Portions:** Many
   restaurant portions are larger
   than necessary. Consider
   sharing a dish with someone or
   asking for a to-go box at the
   beginning of the meal to save
   leftovers for another day.

5. **Be Mindful of Sauces:**
   Restaurant sauces can be high
   in added sugars and unhealthy
   fats. Ask for sauces on the side
   or choose dishes with lighter
   sauce options.

6. **Limit Alcohol:** While
   moderate alcohol consumption
   may have some health benefits,
   excessive alcohol can impair
   brain function. Be mindful of

your alcohol intake when
dining out.

incorporating these practical tips into
your grocery shopping, meal
preparation, and dining-out routines,
you can successfully implement a
brain health diet and support cognitive
function, mood stability, and overall
brain well-being.

CHAPTER 8

Staying Consistent and Overcoming Challenges

Maintaining a brain health diet and staying consistent with your dietary goals can be challenging, but with the right strategies, you can overcome obstacles and achieve long-term success.

8.1 Setting Realistic Goals

Setting realistic and achievable goals is essential for staying consistent with your brain health diet:

- **Start Small:** Begin with manageable changes to your diet. Setting small goals allows you to build confidence and make lasting changes over time.

- **Be Specific:** Define clear and specific goals. Instead of saying, "I want to eat healthier," specify what that means for you, such as "I will eat two servings of vegetables with dinner every night."

- **Prioritize:** Focus on the most important changes first. Identify the dietary modifications that will have the most significant impact on your brain health and prioritize those.

- **Set Deadlines:** Establish timeframes for your goals.

Having a deadline can provide motivation and a sense of urgency.

- **Celebrate Progress:** Celebrate your achievements along the way. Recognize and reward yourself when you reach milestones in your brain health journey.

8.2 Dealing with Cravings

Cravings for unhealthy foods can be a significant challenge. Here's how to manage them:

- **Identify Triggers:** Pay attention to what triggers your cravings. It could be stress, boredom, or specific situations.

- **Healthy Substitutes:** Find healthier alternatives for your cravings. If you're craving something sweet, opt for fruit or dark chocolate with high cocoa content. When craving salty snacks, choose unsalted nuts or whole-grain crackers.

- **Practice Mindfulness:** Use mindful eating techniques to assess whether you're truly hungry or if the craving is emotional. If it's emotional, find alternative ways to cope with stress or boredom.

- **Stay Hydrated:** Sometimes, thirst can be mistaken for hunger. Drink water regularly to stay hydrated, which can help reduce cravings.

- **Plan Ahead:** If you know you have certain trigger foods that lead to overeating or unhealthy choices, try to keep them out of your home or have smaller portions available.

8.3 Managing Social Situations

Eating habits often revolve around social situations. Here's how to manage social settings while sticking to your brain health diet:

- **Communicate Your Goals:** Let friends and family know about your dietary goals so they can support your choices when dining together.

- **Offer to Contribute:** If attending a gathering or

potluck, offer to bring a brain-healthy dish to ensure there's an option that aligns with your diet.

- **Plan Ahead:** Before dining out, check the menu online and choose a restaurant with healthy options. This can help you make better choices when ordering.

- **Stick to Your Values:** Don't feel pressured to indulge in unhealthy foods just because others are. You can politely decline or have a smaller portion.

- **Enjoy the Company:** Focus on the social aspect of gatherings rather than just the food. Engaging in conversations and activities can help distract from food temptations.

8.4 Tracking Your Progress

Tracking your progress is essential for staying consistent and motivated:

- **Food Journal:** Keep a food journal to record what you eat, how you feel, and any cravings you experience. This can help you identify patterns and make adjustments.

- **Use Apps:** There are many smartphone apps available that can help you track your food intake, nutrient consumption, and progress toward your goals.

- **Regular Assessments:**
  Periodically evaluate your
  progress against your goals.
  Are you meeting your targets?
  Do you need to make
  adjustments?

- **Celebrate Achievements:** As
  you reach milestones, celebrate
  your achievements.
  Recognizing your progress can
  boost motivation and reinforce
  your commitment to a brain
  health diet.

- **Seek Support:** Consider
  joining a support group or
  working with a dietitian or
  nutritionist to help you track
  your progress and receive
  guidance and encouragement.

Consistency is key to long-term
success with a brain health diet. Stay

patient, adapt to challenges as they arise, and focus on the positive impact that a healthy diet can have on your cognitive function and overall well-being.

CHAPTER 9

Exercise and Brain Health

Physical activity is closely linked to brain health and cognitive function.

9.1 The Connection Between Physical Activity and Cognitive Function

Exercise has a profound impact on brain health and cognitive function:

- **Improved Blood Flow:** Exercise increases blood flow to the brain, delivering oxygen and essential nutrients that support cognitive processes.

- **Neuroplasticity:** Regular exercise can enhance neuroplasticity, the brain's ability to adapt and change. This can lead to improved learning and memory.

- **Release of Neurotransmitters:** Exercise stimulates the release of neurotransmitters like dopamine and serotonin, which play key roles in mood regulation and cognitive function.

- **Reduced Inflammation:** Physical activity has anti-inflammatory effects, potentially reducing the risk of neurodegenerative diseases associated with chronic inflammation.

- **Stress Reduction:** Exercise is an effective way to reduce stress, which can have a negative impact on cognitive function when chronic.

9.2 Incorporating Exercise into Your Routine

To support brain health through exercise, consider the following tips:

- **Choose Activities You Enjoy:** Engage in physical activities you genuinely enjoy, whether it's walking, dancing, swimming, or playing a sport. Enjoyment makes it more likely

that you'll stick to your exercise
routine.

- **Set Realistic Goals:** Start with
  achievable exercise goals, such
  as walking for 30 minutes most
  days of the week. Gradually
  increase intensity and duration
  as your fitness improves.

- **Mix It Up:** Incorporate a
  variety of exercises into your
  routine. Include aerobic
  exercises like jogging, strength
  training, flexibility exercises
  like yoga, and balance-focused
  activities like tai chi.

- **Schedule Regular Workouts:**
  Make exercise a consistent part
  of your routine by scheduling
  specific times for physical
  activity. Treat exercise
  appointments with the same

importance as other commitments.

- **Prioritize Movement:** Look for opportunities to move more throughout the day, such as taking short walks during breaks, using stairs instead of elevators, or doing stretching exercises.

- **Socialize Through Exercise:** Engage in group fitness classes or invite friends to join you in physical activities. Socializing while exercising can be motivating and enjoyable.

- **Track Your Progress:** Keep a record of your exercise sessions and accomplishments. Tracking your progress can provide motivation and a sense of achievement.

- **Stay Hydrated:** Proper hydration is crucial when engaging in physical activity. Drink water before, during, and after exercise to support both brain and overall health.

- **Consult a Healthcare Professional:** If you have existing medical conditions or concerns about starting an exercise routine, consult a healthcare professional or fitness expert for guidance.

Regular physical activity not only supports brain health but also contributes to overall well-being, helping you maintain cognitive function, reduce stress, and enhance mood and energy levels.

CHAPTER 10

Other Lifestyle Factors

In addition to diet and exercise, other lifestyle factors play a significant role in brain health. Let's explore these factors:

10.1 The Importance of Quality Sleep

Quality sleep is essential for cognitive function and memory consolidation. Aim for 7-9 hours of uninterrupted sleep each night. Establish a consistent sleep schedule, create a relaxing bedtime routine, and ensure your sleep environment is conducive to rest.

10.2 Stress Management Strategies

Chronic stress can negatively impact brain health. Practice stress management techniques such as mindfulness meditation, deep breathing exercises, yoga, or progressive muscle relaxation. Engaging in hobbies and activities you enjoy can also help alleviate stress.

10.3 Social Connections and Brain Health

Maintaining social connections and engaging in meaningful interactions with others is vital for brain health. Strong social ties can reduce the risk of cognitive decline and promote emotional well-being. Make an effort to nurture your relationships and seek out social activities that bring you joy.

These lifestyle factors, including quality sleep, stress management, and social connections, into your brain health plan, you can further enhance cognitive function and overall well-being.

CHAPTER 11

Monitoring and Adapting Your Brain Health Diet

Monitoring and adapting your brain health diet is essential for ensuring that it continues to meet your needs and goals.

11.1 Tracking Your Diet and Its Effects

Monitoring your diet and its effects on your overall health and cognitive function is a valuable practice. Here's how to do it effectively:

- **Keep a Food Journal:** Record what you eat and drink daily, including portion sizes. Note any changes in your energy levels, mood, and cognitive performance.

- **Track Nutrients:** Pay attention to the nutrients you consume. You can use apps or online tools to calculate your daily intake of essential nutrients like omega-3 fatty acids, antioxidants, vitamins, and minerals.

- **Regular Assessments:** Periodically assess how your diet is affecting your health and well-being. Are you experiencing any cognitive improvements or changes in mood? Are there any areas where you could make improvements?

- **Consult a Professional:** Consider consulting a registered dietitian or healthcare provider for a more in-depth assessment of your dietary habits and their impact on your health.

11.2 Adjusting Your Diet as Needed

Adapting your diet over time ensures that it remains effective and sustainable. Here's how to make adjustments:

- **Evaluate Your Goals:** Review your brain health goals and assess whether your current dietary plan aligns with those objectives.

- **Identify Areas for Improvement:** Based on your tracking and assessments, identify specific areas of your diet that may need improvement. This could include increasing certain nutrients or reducing specific foods.

- **Gradual Changes:** Make dietary changes gradually to give your body and taste buds time to adjust. Sudden, drastic changes may be difficult to sustain.

- **Experiment:** Try new brain-boosting foods and recipes to keep your diet interesting and enjoyable. This can help you

stay motivated and committed
to your goals.

- **Consult a Professional:** If
  you're uncertain about how to
  make appropriate adjustments
  to your diet, seek guidance
  from a registered dietitian or
  healthcare provider. They can
  offer personalized
  recommendations based on
  your needs and goals.

- **Consider Supplements:** In
  some cases, dietary
  supplements may be
  recommended to fill nutrient
  gaps. Consult a healthcare
  professional before taking
  supplements to ensure they are
  appropriate for your specific
  situation.

- **Stay Informed:** Continue to educate yourself about the latest research on nutrition and brain health. Stay open to adapting your diet based on new findings and recommendations.

Dietary needs can vary from person to person, so what works for one individual may not work for another. It's essential to tailor your brain health diet to your specific goals, preferences, and any unique dietary requirements you may have.

monitoring your diet, regularly assessing its effects, and making adjustments as needed, you can ensure that your brain health diet remains effective in supporting cognitive function, mood stability, and overall well-being.